Daytime
and
Nighttime
Animals

by Barbara J. Behm
Illustrated by Martin Camm

Gareth Stevens Publishing
MILWAUKEE

For a free color catalog describing Gareth Stevens Publishing's list of high-quality books and multimedia programs, call 1-800-542-2595 (USA) or 1-800-461-9120 (Canada). Gareth Stevens Publishing's Fax: (414) 225-0377.

Library of Congress Cataloging-in-Publication Data

Behm, Barbara J., 1952-
    Daytime and nighttime animals / by Barbara J. Behm; illustrated by Martin Camm.
       p. cm. — (Animal opposites)
    Includes index.
    Summary: Introduces ten animals and their behavior, including the lemur, marine iguana, and fox, identifying them either as daytime or nighttime animals.
    ISBN 0-8368-2459-8 (lib. bdg.)
    1. Animals—Miscellanea—Juvenile literature. [1. Animals. 2. Nocturnal animals.]
  I. Camm, Martin, ill. II. Title. III. Series: Animal opposites (Milwaukee, Wis.)
  QL49.B364   1999
  590—dc21   99-32254

This North American edition first published in 1999 by
Gareth Stevens Publishing
1555 North RiverCenter Drive, Suite 201
Milwaukee, WI 53212 USA

This edition © 1999 by Gareth Stevens, Inc. Created with original © 1997 by Horus Editions Limited, a division of Award Publications Limited, 1st Floor, 27 Longford Street, London NW1 3DZ, U.K. Additional end matter © 1999 by Gareth Stevens, Inc.

Cover illustrations: a screech owl and a chimpanzee

Printed in the United States of America

1 2 3 4 5 6 7 8 9 03 02 01 00 99

Contents

Lemur

The lemur is a daytime animal.

The lemur searches for food during much of each day. It eats fruit, leaves, and insects.

Baby lemurs
ride on
the back of
their mothers.

Genet

The genet is a nighttime animal.

The genet sleeps during the day. After dark, it comes out to hunt small animals.

Genets
are thin
and look a
little like cats.

7

Pelican

The pelican is a daytime animal.

The pelican spends the day fishing. It dips its head under the water and scoops up prey.

Pelicans
have long beaks.

Owl

The owl is a nighttime animal.

The owl sees very well in the dark. It can hear quiet sounds. It is a bird of prey.

Some owls
look like
tree bark.

Chimpanzee

*The chimpanzee
is a daytime animal.*

The chimpanzee lives in Africa. It makes tools and uses them to find food in the daytime.

Chimpanzees use twigs
to collect ants to eat.

Badger

The badger is a nighttime animal.

The badger spends daytime hours underground. It digs deep burrows in hillsides.

Badgers come out at
night to hunt for food.

Marine Iguana

The marine iguana
is a daytime animal.

The marine iguana suns itself on rocks each day. It dives into water to eat seaweed.

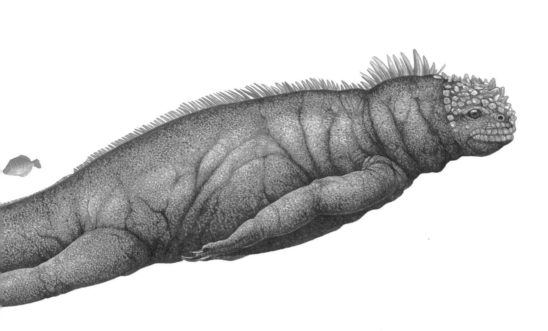

Marine iguanas are the
only type of lizard that
swims in the sea.

Fox

The fox is a nighttime animal.

The fox lives in towns and the countryside. It sleeps in a den underground during the day.

Foxes pounce
on their meals.

19

Kangaroo

The kangaroo is a daytime animal.

The kangaroo runs fast. It jumps high with its long legs. It balances with its tail.

Kangaroo
mothers carry their
babies in a pouch.

Bat

The bat is a nighttime animal.

The bat flies at night but cannot see in the dark. It finds objects and food using echoes.

Vampire
bats feed
on the blood
of other animals.

Glossary

burrow: a hole in the ground that shelters animals.

echoes: sounds that bounce off various surfaces.

marine: living or growing in the sea.

pouch: part of an animal's body that is like a bag.
  A kangaroo mother carries her young in her pouch.

prey: any animal that is hunted by another animal.
  A bird of prey hunts other animals.

Index